```
     E                           2531
  567.9        Gans              $9.00

Water for Dinosaurs & You
     DATE DUE
```

IMPERIAL PUBLIC LIBRARY
P.O. BOX 307
IMPERIAL, TEXAS 79743

Water for Dinosaurs and You

Everywhere on earth there is water—in lakes and rivers and oceans, in the air and under the ground. There are giant icebergs and sparkling drifts of snow. All the water that was on the earth in the beginning is still here today, every drop of it, and people use it over and over again. It will always be here, but there will never be any more, so we must learn to use it wisely.

    Simply and clearly Roma Gans explains how water goes from earth to air, to clouds, to rain, and back to earth again. She tells what happens as farm and factory wastes pollute the lakes and rivers, and what men are doing to protect our essential water supply. From this easy-to-read book, young ecologists will learn to understand and appreciate the importance of our most valuable natural resource.

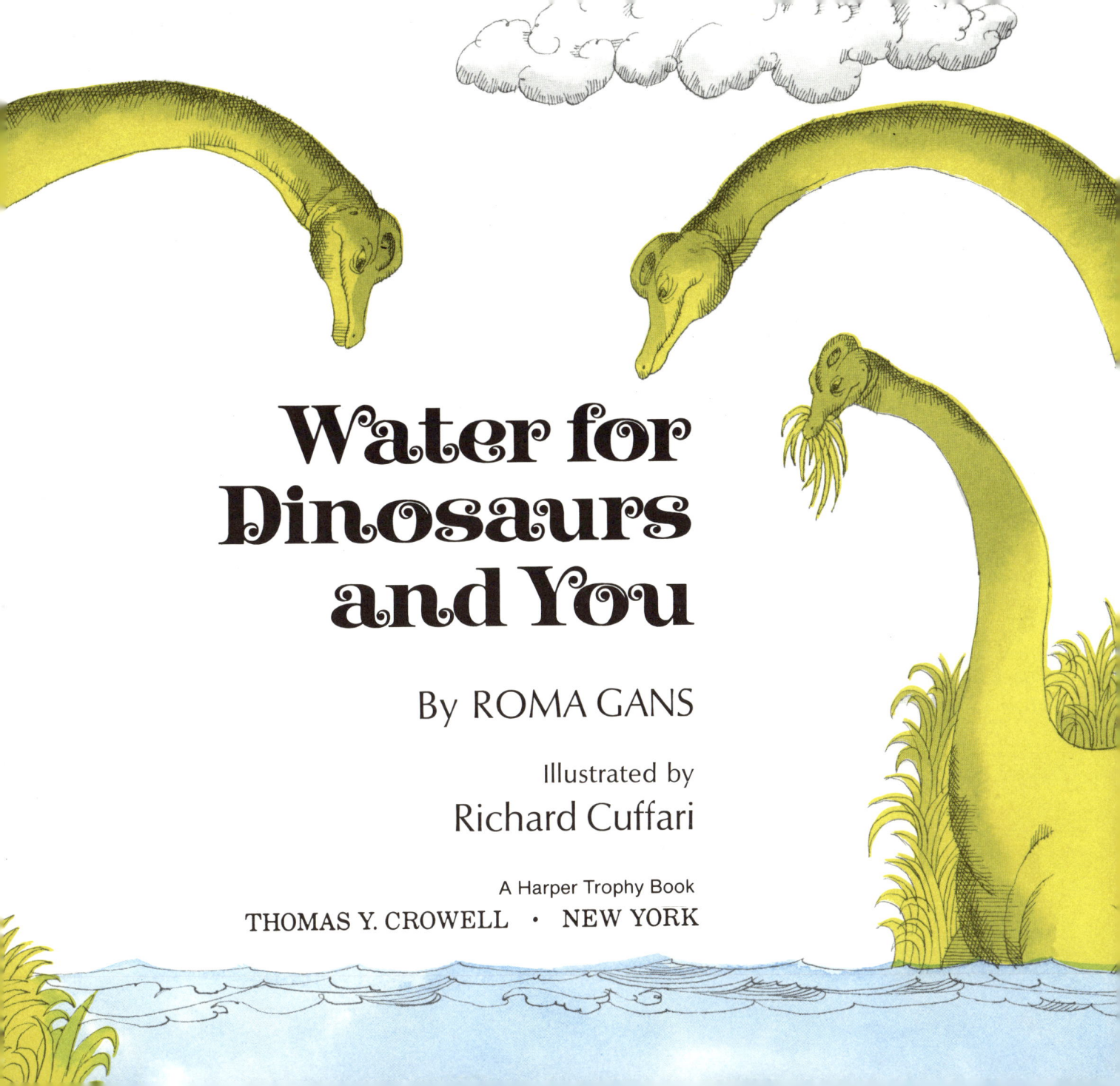

Water for Dinosaurs and You

By ROMA GANS

Illustrated by Richard Cuffari

A Harper Trophy Book
THOMAS Y. CROWELL · NEW YORK

LET'S-READ-AND-FIND-OUT BOOKS

Let's-Read-and-Find-Out Books are edited by Dr. Roma Gans, Professor Emeritus of Childhood Education, Teachers College, Columbia University, and Dr. Franklyn M. Branley, Astronomer Emeritus and former Chairman of the American Museum–Hayden Planetarium. Text and illustrations for each of the more than 100 books in the series are checked for accuracy by an expert in the relevant field. Other titles available in paperback are listed below. Look for them at your local bookstore or library.

A Baby Starts to Grow
Bees and Beelines
Birds at Night
Corn Is Maize
Digging Up Dinosaurs
A Drop of Blood
Ducks Don't Get Wet
Fireflies in the Night
Follow Your Nose
Fossils Tell of Long Ago
Hear Your Heart
High Sounds, Low Sounds
How a Seed Grows

How Many Teeth?
How You Talk
It's Nesting Time
Ladybug, Ladybug, Fly Away Home
Look at Your Eyes
Me and My Family Tree
My Five Senses
My Visit to the Dinosaurs
No Measles, No Mumps for Me
Oxygen Keeps You Alive
The Planets in Our Solar System
The Skeleton Inside You
The Sky Is Full of Stars

Spider Silk
Straight Hair, Curly Hair
A Tree Is a Plant
Water for Dinosaurs and You
Wild and Woolly Mammoths
What Happens to a Hamburger
What I Like About Toads
What Makes Day and Night
What the Moon Is Like
Why Frogs Are Wet
Your Skin and Mine

Copyright © 1972 by Roma Gans. Illustrations copyright © 1972 by Richard Cuffari.
All rights reserved. Published simultaneously in Canada by Fitzhenry & Whiteside Limited, Toronto.
Manufactured in the United States of America.   L.C. Card 78-158691   ISBN 0-690-00202-5

Water for Dinosaurs and You

Today, you may drink some of the same water that a dinosaur drank millions of years ago.

The water may have come from a deep well, or a lake, or a river.
It looks clear.
It tastes fresh.
But this water you drink today has been on earth millions and millions of years.
It has been used over and over again.

Some has been drunk by animals.

Some has watered a king's garden.

Some has been used to bathe babies in faraway countries.

The water in a glass looks clear and clean.
It even sparkles.
But it is the same old water that has been on earth for millions of years.
It may sparkle in the glass.
Yet it is used water.

All the water that was on earth in the beginning is here today, every drop of it. But there will never be any more water than we have now.

There are thousands and thousands of lakes full of water for us to use.

There are millions of miles of rivers, too.

High in the mountains there is snow and there are glaciers. This is frozen water. When the frozen water melts, rivers carry it down into valleys and plains.

All of this is water that we can see.

There is more water that we cannot see.
It is underground. We call it groundwater.
Groundwater is under cities.
It is under the roads we ride on.
It is even under deserts.
There is water somewhere beneath you, no matter where you are right now.

More water is underground than in all our thousands of lakes and millions of miles of rivers.
Much of it is in soil, sand, and rocks.
The soil, sand, and rocks hold the water the way a sponge does.

There is water in the air, too.
You can feel it on a damp day.
You can really see it when you look at the clouds above you.
The water in the air, in the soil under the ground, in lakes, and in rivers is called fresh water.

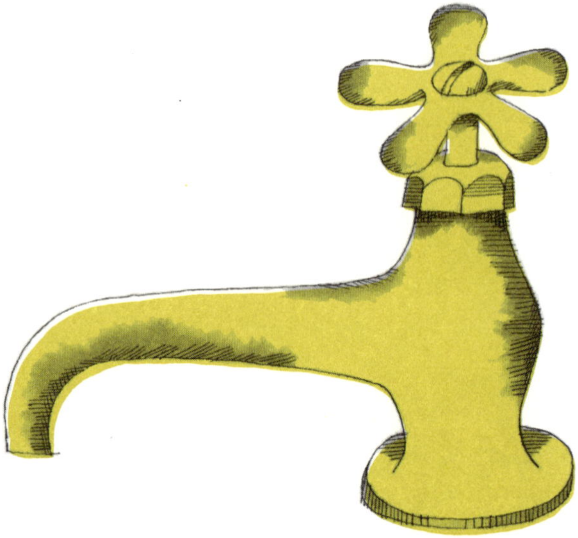

We use fresh water in our houses, stores, and factories.
We get fresh water from lakes and rivers.
And we get it from wells. The well water is pumped
   from underground.

Even more water is in the oceans than is in our lakes
  and rivers and underground.
Ocean water is salt water.
It is too salty to drink.
It is too salty to use in our houses.

No matter how much water we use in our houses,
   lakes are still full of water.
And rivers keep running.
The oceans, too, stay the same.
So it has been for millions of years.

How does this happen?

Some water goes into the air. When it does this, we say it evaporates.

The water becomes a vapor.

You cannot see water vapor, but on a damp day you can feel it.

You say you feel sticky.

The vapor changes into tiny droplets.
Billions of tiny droplets make clouds.
The water in the droplets may have come from the Atlantic or the Pacific Ocean.
It may have come from a swimming pool in your town.

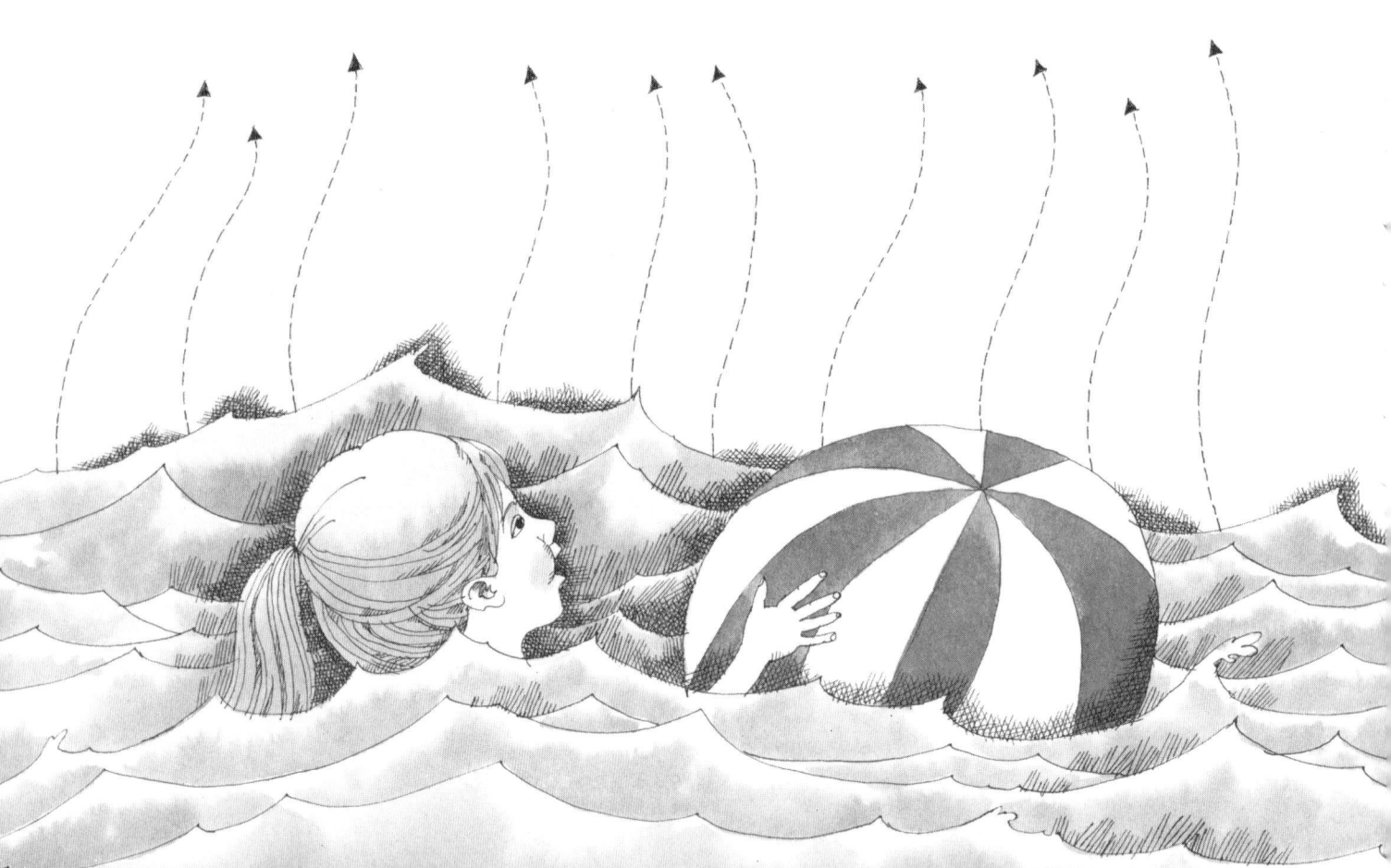

The droplets in clouds come together to make big drops. Then these drops fall back to the earth. We say, "It's raining."

Some of the rain falls into lakes, rivers, and oceans. Some falls on soil and slowly seeps into the ground. This will become groundwater.

Some rushes along in rivers until it runs into the ocean.

As water flows along in a river, it washes minerals and salt from soil and rocks.
The river carries the minerals and salt to the ocean.
Rivers have been carrying minerals and salt to the oceans for millions of years.
So the oceans get saltier and saltier.
Ocean water is salt water, not fresh water.

We need fresh water to use in houses, stores, and factories.

Rainwater is fresh water, and when it starts out from the clouds it is clean water.

When it falls as snow on the icecaps of the North or South Pole, it does not pick up any soot or dust.

It forms ice. When this ice melts it becomes clean water.

But rain picks up soot and dust as it falls through the air over towns and cities. Then it is not clean.

The water in our wells has salt and minerals in it. So it is not really clean water either, even if it is good to drink.

Water that has been used in houses, stores, and factories has soap, dirt, and other wastes in it. We say it is polluted.

The rain that runs off a farmer's fields carries soil and pesticides with it.
The soil and pesticides are washed into lakes and rivers.
They, too, pollute the water.

Some water is too polluted for us to swim in.
Our hands would get dirty if we washed them in some lakes.
Fish cannot live in polluted water.
Water in some rivers is too polluted even for factories to use.

Oxygen is in all lakes and rivers.
The oxygen helps clean the water and keep it fresh.
When too much waste gets into water, even oxygen cannot clean it.

Cities near polluted lakes and rivers must clean the water they use.
But it takes a long time to clean water, and it costs a lot of money.
We must learn how to use water without polluting it.
Many farmers have stopped using pesticides that would pollute water.
Many factories have stopped flushing waste into rivers.

It may take many years to clean some lakes and rivers.
They are very polluted.
Once they are clean again, their water can be used over and over.

The water will keep on going—from earth to air, to clouds, to rain, and then back to earth again.
It will fall from the clouds.
It will fill lakes and rivers and ponds.
It will fill oceans.
All the water will be used water.
And all the water will be millions of years old.

ABOUT THE AUTHOR

Roma Gans has called children "enlightened, excited citizens." She believes in the fundamental theory that children are eager to learn and will whet their own intellectual curiosity if they are encouraged by and provided with stimulating teachers and books.

Dr. Gans received her B.S. from Columbia Teachers College and her Ph.D. from Columbia University. She began her work in the educational field in the public schools of the Middle West as a teacher, supervisor, and assistant superintendent of schools. She is Professor Emeritus of Childhood Education at Teachers College, Columbia University, and lectures extensively throughout this country and Canada.

Dr. Gans lives in West Redding, Connecticut, where she enjoys observing the many aspects of nature.

ABOUT THE ILLUSTRATOR

Richard Cuffari is a painter and an illustrator of children's books. He has received awards from the American Institute of Graphic Arts and the Society of Illustrators.

A native of New York, Mr. Cuffari studied at Pratt Institute. He lives in Brooklyn with his wife and four children.

IMPERIAL PUBLIC LIBRARY
P.O. BOX 307
IMPERIAL, TEXAS 79743